The Poultry Industry in Oregon

by Professor James Dryden

with an introduction by Jackson Chambers

Self Reliance Books

Get more historic titles on animal and stock breeding, gardening and old fashioned skills by visiting us at:

http://selfreliancebooks.blogspot.com/

Introduction

I am pleased to present yet another title on Poultry.

The work is in the Public Domain and is re-printed here in accordance with Federal Laws.

As with all reprinted books of this age that are intended to perfectly reproduce the original edition, considerable pains and effort had to be undertaken to correct fading and sometimes outright damage to existing proofs of this title. At times, this task is quite monumental, requiring an almost total "rebuilding" of some pages from digital proofs of multiple copies. Despite this, imperfections still sometimes exist in the final proof and may detract from the visual appearance of the text.

I hope you enjoy reading this book as much as I enjoyed making it available to readers again.

Jackson Chambers

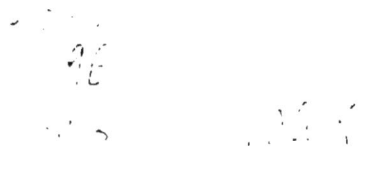

Fig. 1—View on the poultry farm of H. Ringhouse, Gresham, Oregon.

Fig. 2.—Rear view of poultry houses. Mr. Ringhouse's farm, showing double yards.

Fig. 3—View on Mr. Joseph Schulte's farm at Sublimity, Oregon.

Fig. 4 — A poultry farm at Petaluma, California, showing colony system of housing. Note houses scattered over the hills.

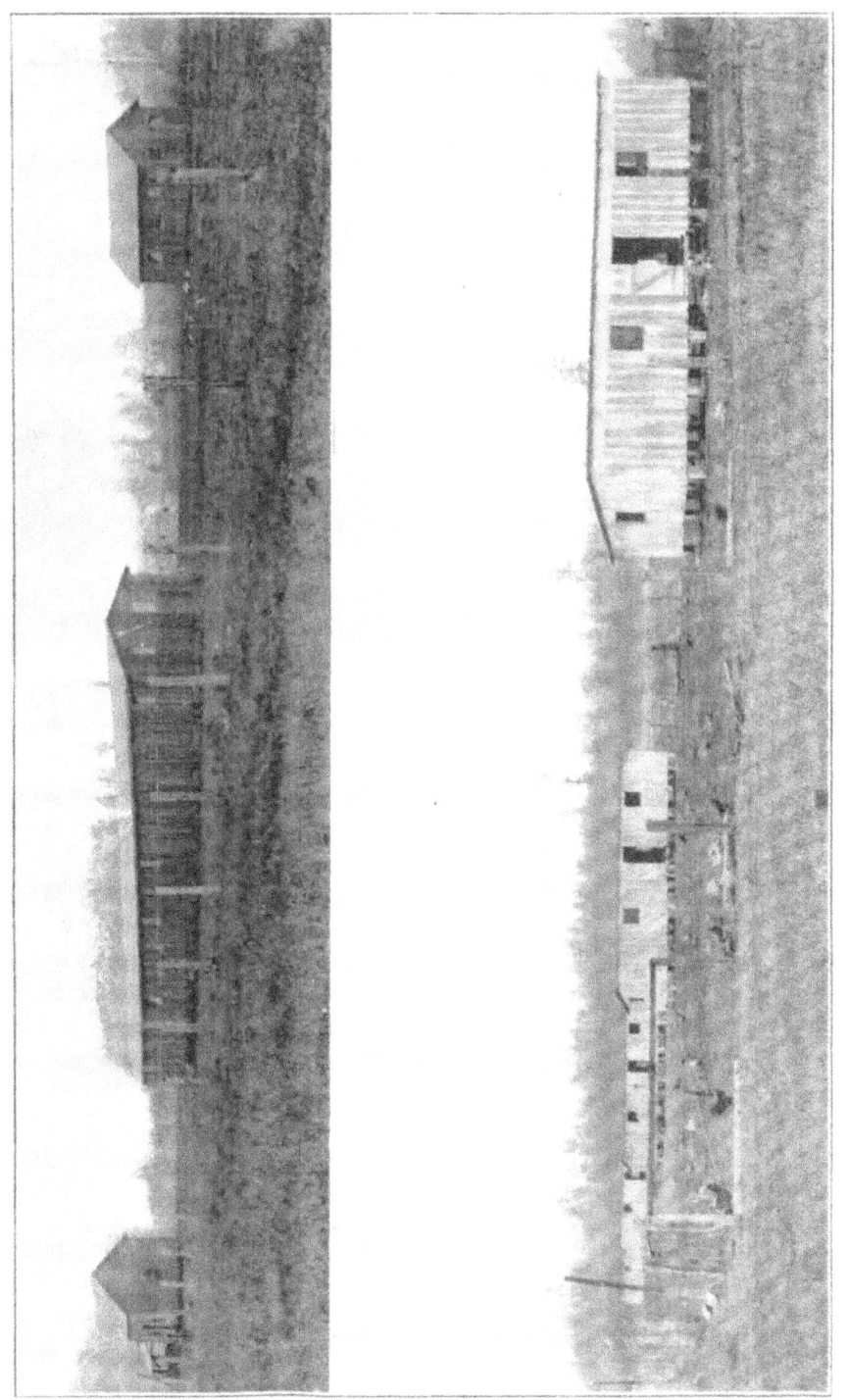

Fig. 5—Two views of farm of Mr. J. Lenz, Hillsboro, Oregon.

Fig. 6—Colony house on poultry farm of Mr. E. K. Brown, Eugene, Oregon.

Fig. 7—Colony house made out of two piano boxes on farm of E. K. Brown, Eugene, Oregon.

Fig. 8—An open front stationary house belonging to Mr. H. S. Poisel, Salem, Oregon.

Fig. 9—Car for shipping live poultry.

Fig. 10—The shingled fresh-air house on Mr. Ringhouse's farm at Gresham, Oregon.

Fig. 11—Poultry invading the forest. A year ago Mr. Ringhouse's poultry farm was covered with timber

Fig. 12—A house 800 feet long on farm of Mr. F. L. Miller, Corvallis, Oregon, being taken down after being proven unsatisfactory.

Fig. 13—The style of house now used with satisfactory results by Mr. Miller, of Corvallis, on his Buff Orpington farm.

Fig. 14—A stationary house on the Ladd farm, near Portland.

Fig. 15—A movable brooder house on the Ladd farm.

Fig. 16—A movable open-front house on the Krebs hop farm at Brooks, Oregon.

Fig. 17—View showing one of the laying houses belonging to William Rogers, Grants Pass, Oregon, and a fence without postholes.

Fig. 18—A movable colony house with curtain-front, at the Utah Experiment Station.

Fig. 19—View showing a combination of poultry and fruit.

Fig. 20—Illustrating the three-house colony system in California, and showing a feeding trough for mash.

Fig. 21—Another type of California colony house.

Fig. 22—View on a California poultry farm, showing colonies widely separated.

Fig. 23—Another California colony poultry farm.

THE POULTRY INDUSTRY IN OREGON.

By JAMES DRYDEN.

INTRODUCTORY.

.This bulletin has been prepared with the object of encouraging the development of the poultry industry in Oregon. It is the purpose to publish other bulletins in the future dealing with different phases of the industry. The writer during the few months he has been in the State has made diligent inquiry into the conditions of the industry and has visited most of the poultry producing sections of the Willamette Valley. His observations so far as they have extended are given in the following pages, together with suggestions looking toward improvement, which it is hoped may be helpful to those now interested in poultry-keeping as well as to others who contemplate engaging in the business.

THE IMPORTANCE OF THE INDUSTRY.

The poultry industry in Oregon is an important one. Its importance can be fully understood only by a study of the statistics of production, and the statistics, unfortunately, are not very complete. This State has collected no statistics of poultry products. The only reliable data we have is contained in the last U. S. Census made in the year 1899. This census gives the number of chickens, turkeys, ducks and geese, over three months of age, on June 1, 1900, the value of the poultry raised in 1899, the number of dozens of eggs produced and the value of the same. This information is given for each county in the State in accompanying table.

The poultry and the eggs produced in villages, towns and cities were not included in the enumeration. It should also be stated that the table gives the number of fowls, over three months of age only on June 1st. This means that very little was counted except breeding fowls. Eggs were valued at an average of 15 cents per dozen. In total value of poultry and eggs Linn County stood first, followed by Marion, Clackamas, Lane, Douglas, Washington, and Yamhill in the order named.

U. S. Census of Poultry and Eggs in Oregon.

	Chickens, including Guinea fowls	Turkeys	Geese	Ducks	Value of all poultry June 1 1900	Value of poultry raised in 1899	Dozens of eggs produced in 1899
The State	1,290,818	36,031	26,580	19,774	$582,524	$826,687	7,709,970
Baker	28,153	630	684	427	12,931	20,527	148,430
Benton	34,349	1,023	1,089	266	18,024	24,640	223,250
Clackamas	84,919	1,229	1,871	696	46,551	59,248	659,310
Clatsop	8,983	24	41	170	4,793	5,877	68,820
Columbia	18,101	139	244	180	8,624	11,657	123,020
Coos	23,847	356	828	586	8,663	11,631	165,820
Crook	14,001	360	19	182	5,395	8,031	68,860
Curry	7,911	142	178	204	2,420	3,248	32,830
Douglas	67,238	8,111	1,733	557	41,002	78,740	368,410
Gilliam	14,462	453	68	179	5,865	9,322	108,020
Grant	18,829	277	270	156	6,082	9,643	91,940
Harney	6,536	47	20	1	2,442	4,081	43,760
Jackson	55,818	3,585	393	555	20,599	31,989	301,310
Josephine	15,596	261	72	147	4,597	8,078	91,840
Klamath	9,605	191	7	7	4,590	4,850	51,270
Lake	8,198	29	40	3	2,663	4,400	39,410
Lane	97,331	3,623	2,601	1,192	41,657	65,217	514,520
Lincoln	7,992	151	414	80	3,485	4,125	55,640
Linn	107,358	3,168	4,101	2,233	59,585	83,368	670,420
Malheur	16,174	299	64	199	8,196	11,287	112,550
Marion	114,748	1,473	3,460	2,036	49,612	64,160	782,170
Morrow	22,771	1,078	102	440	8,045	13,200	148,590
Multnomah	51,753	515	394	3,502	22,433	25,909	322,810
Polk	47,721	911	1,436	611	22,727	34,643	263,210
Sherman	20,752	528	829	356	8,472	13,187	122,650
Tillamook	13,062	115	256	205	4,182	6,199	90,400
Umatilla	71,363	1,514	899	1,117	31,435	46,476	363,070
Union	59,864	877	768	812	28,205	37,191	355,290
Wallowa	31,999	449	218	275	9,956	13,234	111,160
Wasco	52,171	1,836	256	609	19,244	20,463	211,330
Washington	79,207	551	1,237	803	34,773	42,521	533,140
Wheeler	10,480	211	14	176	3,738	3,936	41,900
Yamhill	63,700	1,672	1,891	663	29,965	42,627	458,690
Grande Ronde ₁	1,598	2	36	3	508	691	6,030
Klamath ₁	473	6	1		132	156	2,270
Siletz ₁	600		11	2	166	166	1,090
Umatilla ₁	3,155	195	35	144	1,397	1,974	11,740

₁ Indian reservation.

It is of interest to compare the poultry production of Oregon with that of other states. The census for 1899 reports a production of eggs in the United States of 1,299,819,186 dozen, of a value of $144,286,151, on a price basis of 11.1 cents per dozen. The poultry produced on the farms during the same year was estimated at $136,891,877. If the amount produced on the town, city or village lot, were added, it would increase the total very much. The north central division of states, including the Mississippi Valley, is the greatest producing section in the United States, showing that the region of greatest agricultural development is the region of greatest poultry production. Comparing the different states, we find Iowa was the greatest producer of eggs but the eggs produced in Ohio had slightly the greatest value on account of higher prices. Each of these states produced eggs in 1899 worth about $10,000,000. In value of poultry raised Illinois led with over $11,000,000 worth. The farms of Iowa, Illinois and Ohio produced about $60,000,000 worth of poultry and eggs in that year, From the report of the State Labor Bureau of Missouri, the poultry products of that state for 1906 were valued at $39,725,539. It has been estimated that the states of Iowa, Missouri, Illinois and Kansas produced over twenty-five per cent. of the poultry and eggs marketed in the United States in 1899.

It is instructive to compare these values with those of some of the other industries of the country. During the census year the oat crop lacked seventy-four million dollars of being equal to the the poultry crop; barley had a value of just one-seventh that of poultry and eggs; all the orchards of the United States produced less than one-third as much as the poultry; fruit of all kinds was worth less than one-half as much as the hen crop; and potatoes one-third as much. The sugar beet crop, which has been the beneficiary of legislation and bounties, does not begin to compare in value with the products of the poultry yards, which worry along with no subsidy but neglect; the wool clip of the United States, a frequent factor in political campaigns and of congressional debates, was worth $45,723,739 in the census year, or $235,454,508 less than the poultry products.

In circular No. 73 of the Bureau of Animal Industry, U. S. Department of Agriculture, the following appears:

"Another statement that will, no doubt, come as a surprise to most people is that the egg product of 1899 was valued at a higher figure than the combined gold and silver product of the United States during any year since 1850, except for the one year of 1900, when the precious metals exceeded the eggs by $9,-418,125. The same statement is true of the poultry product, if we except the years 1899 and 1900, when the excesses of gold and silver combined over eggs were, respectfully, $4,967,123 and $16,812,618. The surprise occasioned by these figures is still further heightened when we become aware that the poultry and eggs together in 1900 were worth more than either the gold or the silver production of the world for any year since the beginning of records, in 1493, excepting the two years of 1891 and 1899, when the poultry products fell below to the extent of $5,701,453 and $25,990,553, respectively."

In spite of the splendid showing for the poultry, the figures do not do the industry full justice. The value represents the amount produced on the farms only and was based on prices that were probably twenty-five per cent. lower than prices are today. It is well known that the industry is not confined to the farms alone, although admittedly the large proportion of the product comes from the farms under a system of mixed husbandry, a system on which the agriculture of this country was wisely founded. Were the business of poultry-keeping confined to the poultry farms this system would lose an important support and at the same time the cost of producing poultry and eggs would undoubtedly be greater. So long as the system of mixed husbandry prevails and the poultry products are largely produced on the by-products of that system of farming, we may expect the large share of the poultry products to come from the farm.

But considerable numbers of poultry are kept in the suburbs of cities and on town and village lots, either as a diversion or as a certain means of securing eggs guaranteed strictly fresh. Were these amounts added to the census figures, and were the prices that prevail today used in computing the values those values would be very much greater.

As stated above, no census has been taken of the agricultural products since 1899 and it is impossible to show accurately what the development has been since that year, but the United States Department of Agriculture maintains a statistical bureau whose business it is to gather annual statistics of crops and animal products. These are published every year. The report of Secretary Wilson for 1907, submitted to the present Congress of the United States, places the value of the poultry products of the United States at the stupendous amount of $600,000,000, nearly $50,000,000 more than the value of the wheat crop for the same year.

The value of the poultry industry in the State of Oregon is compared with the other industries of the farm, in the following table compiled from the census report for 1899:

Animals sold	$6,598,323
Wheat	6,358,395
Hay and forage	6,147,018
Milk, butter and cheese	3,550,953
Wool	2,396,741
Oats	2,078,950
Poultry and eggs	1,988,758
Animals slaughtered	1,565,895
Hops (10 cents per pound)	1,467,557
Fruit of all kinds	1,455,201

To show how the poultry business has been growing in this State I quote the following from the census reports for three decades:

1879................... 1,654,738 dozens of eggs
1888................... 4,453,933 " "
1899................... 7,709,970 " "

Basing an estimate on the three previous decades, including the amounts other than those produced on the farms, and allowing for the higher prices of the present time, it is safe to say that the value of the poultry products in 1907 will reach a total of four million dollars. The importance of the industry would be better realized if poultry-keeping were to be abandoned and this sum of money sent out of the State for poultry and eggs.

The industry is therefore an important one in this State, and when it is considered that these results are secured without any particular effort to get the best possible returns from the poultry, they are remarkable. No other branch of agriculture probably has been more neglected than this, and little has been done in a systematic way to increase the production of the flock on the average farm. This is shown by the fact that the hens of Oregon were credited by the census with an average of only 72 eggs per hen when it is possible to nearly double that yield by better methods and better breeding.

THE MARKET.

The market conditions are favorable for increased production of poultry and eggs in this state. In few sections of the country are the prices of poultry and eggs better than they are in Oregon and on the Pacific Coast generally. At the present time (December) fresh ranch eggs are quoted in Portland at 40 cents per dozen wholesale, while consumers are paying as high as 50 cents. At such a price a dozen eggs will about pay for the cost of food consumed by the hen in a year when kept on a farm. The market quotations at Seattle on the north and San Francisco on the south are higher than at Portland and these markets would take care of any possible surplus that might be produced in Oregon.

Importations.—That the markets here are better than in other sections of the country is shown by the fact that large quantities of eggs and poultry are imported from the middle west states. The writer has endeavored to secure reliable information as to the quantity of eggs that are shipped into this State and has been only measurably successful. Commission men and others have been interviewed and it can almost certainly be stated that 75 cars of eggs came into Oregon in 1907, from other states, and there is a probability that the number reached about 100 cars. A car contains 400 cases of eggs, 30 dozen in each case, making 30,000 cases in 75 carloads. These shipments come in in the fall and winter when eggs are at the highest price, but they are usually storage eggs and do not, of course, command the price of local ranch eggs. Figuring on a basis of 75 cars of 30,000 cases at 25 cents a dozen, there was sent out of the State in 1907 $225,000 for eggs.

The information as to the importations of poultry is not as complete as I would like it to be. Estimates made by commission men and others vary from 5 cars to about 20 cars of dressed poultry. About the best information obtained indicates that the product reached a value of from $75,000 to $100,000 last year. The state more nearly meets the demand for poultry than for eggs.

But the local supply has not been keeping pace with the demand for poultry. Mr. Young of Oakland, one of the largest shippers of turkeys in the State, informs me that formerly Portland could not take all the turkeys that were produced at Oakland and a large proportion of the shipments went to San Francisco, but during the last two or three years Portland has been able to take all that could be shipped from that point, although the production has been greatly increased. Very few cars of poultry were brought into Portland from other states until within the last year or two.

The demand for both poultry and eggs has been increasing faster than the supply and there is no danger of overproduction. The reverse is true. Not only does the City of Portland depend on eastern eggs to a large extent during the winter but it is a fact that many of the small towns in the rural communities of the State are buying eastern eggs from Portland dealers. It is a condition that is undesirable that the poultry "farmers" of Front Street, Portland, are selling eggs to the poultry raising communities of the Willamette Valley. This may probably be excused on the theory that these communities prefer to use the cheaper eastern product than the high priced eggs that their own hens lay in the winter. But the shipments of ranch eggs to Portland in winter are very light.

Markets in Neighboring States.—Not only is there a good market in Oregon for eggs and poultry there are also good markets in neighboring states. Eggs that are quoted in Portland at 40 cents are quoted in Seattle at 48 cents and in San Francisco at 55 cents. Our markets therefore will be governed largely by those to the north and south of us. Those markets will take any possible surplus we may have. The importations of eggs and poultry into Seattle are very much greater than into Portland. According to a published statement of Mr. J. L. Anderson, Superintendent of the poultry department of the Alaska-Yukon Exposition, Seattle imported in 1906 250 cars of eggs and 155 cars of poultry, valued at $810,000, figuring eggs at an average of 29½ cents a dozen and poultry at 14 cents per pound.

I have not the figures showing the importations into San Francisco but they reach a tremendous sum. The shipments of live poultry from the East to San Francisco are large. A special poultry car is used for these shipments which holds from 3000 to 5000 chickens, according to size, two such cars from Kansas being unloaded the day I was there. It was learned from the commission dealers there that they could get better live poultry from the East than from Oregon and for that reason very little if any Oregon poultry was going to San Francisco. It was stated that eastern poultry came there in better flesh and it was explained that this was because the Oregon poultry was held too long before being shipped. Where a dealer has to hold chickens several days or a week, as is the case in Oregon, before he can get enough to fill a car, the chickens lose flesh and by the time they get to their destination they are very unsatisfactory. This could be remedied, however, by cooperation among the farmers, as well as by a larger production of good chickens.

The prices of ranch eggs and of eastern eggs in Portland for 1906 and 1907 are shown in the accompanying table which has been compiled from the market reports of The Oregonian. For the best grade it will be noticed that the lowest quotation during 1906 was in March and April when it was 16 cents a dozen for a few days, and the highest price obtained during that year was 37½ cents which was in December. During 1907 the lowest price quoted was 17½ cents which was the quotation for about a week in March, the highest being 40 cents in January, February, November and December.

Prices of "Ranch Eggs," Portland, Oregon.

1906	Jan.	Feb.	March	April	May	June	July	Aug.	Sept.	Oct.	Nov.	Dec.
1	----------	24-25	16-17	15 1-2-16	17½-18	19-19 1-2	22-22 1-2	21-21½	24-25	29-30	32 1-2-35	35-37 1-2
2	----------	24	16-16½	"	18	"	"	"	"	30	"	"
3	28-30	23½-24	15½-16½	16	"	"	"	"	"	"	33-35	"
4	"	23-24	16	"	"	"	"	"	"	"	"	"
5	29-30	"	"	"	18-18½	19-20	"	"	25	30-31	"	"
6	28-30	"	15½-16	"	"	19½-20	21½-22	"	25-26	"	"	"
7	"	"	"	"	"	"	"	21	"	"	"	37½
8	"	"	"	"	"	"	"	"	26-26½	"	"	35-37½
9	29-30	"	15-16	"	18½-19	20	"	21-21½	"	31-32	"	"
10	28-30	23-23½	"	16-16½	"	"	22-22½	"	"	"	"	"
11	28-29	"	"	"	"	"	22½	"	"	"	"	"
12	"	"	"	16½-17	"	20-21	22	"	"	"	"	"
13	"	22½-23	15½-16	17	"	"	21½-22	"	26½-27	31-32½	"	"
14	"	20-22½	16½-17	"	"	"	"	21-22	"	"	"	"
15	"	19-20	"	16½-17	19-20	20-21½	"	"	"	"	"	"
16	27½-29	18-19	16½	"	"	21-22	"	"	"	"	35	"
17	27½-28	17-18	"	"	"	"	"	"	"	"	"	"
18	"	16-17½	"	"	"	"	"	"	27-27½	"	"	35
19	"	"	"	"	"	"	21-22	"	"	"	"	"
20	27½-30	16-17	16-16½	"	"	"	"	"	27½	"	"	"
21	27½-28	16-16½	"	"	"	"	"	21½-22	27½-28	"	35-36	32½-35
22	"	16$17	"	"	"	"	"	"	27½-28½	"	"	"
23	"	16-16½	"	"	"	21½-22	"	"	"	"	35-37½	"
24	"	"	16	17	"	"	21-21½	20½-22	"	"	"	"
25	"	16-17	"	17-17½	"	"	"	22	29	"	"	"
26	27-27½	"	"	"	"	"	21	"	"	32-32½	"	"
27	26-27½	"	"	"	"	"	"	"	"	"	"	30-32
28	26-26½	16½-17	"	"	19-19½	22-22½	"	23-23½	29-30	"	"	30-32½
29	"	----------	"	17½	"	"	"	23½-24	"	"	"	"
30	25-26	----------	15½-16	"	"	"	"	"	"	32½-35	"	"
31	----------	----------	----------	"	"	"	"	"	"	"	"	"

Prices of "Ranch Eggs," Portland, Oregon.—*Continued.*

1907	Jan.	Feb.	March	April	May	June	July	Aug.	Sept.	Oct.	Nov.	Dec.
1		87½	18-19	22-22½	18-18½	17½-18½	17½-18	21½-22	26-27	32½		
2		"	18	19-20	"	"	"	22-22½	"			
3		"	"	18-19	"	"	"	"	"			
4	30-32	"	"	"	"	"	"	"	27-28			
5	"	40	17-18	18	"	18-18½	"	"	27½-28			
6	"	"	17-17½	18½-19½	"	"	"	22-24	"			
7	"	35-37½	17	"	18	"	"	"	"			
8	30-32½	32½-35	17-17½	"	"	"	"	"	"			
9	32-32½	30	"	19-20	"	"	"	"	"			
10	32½-33	"	"	"	"	"	"	"	28			
11	"	"	"	"	"	18½-20	"	"	28-30			
12	"	"	"	19½-20	"	"	"	"	"			
13	"	24-25	"	20	"	"	23-25	23 24	"			
14	"	23-25	17½-18	"	"	20	"	"	"			
15	35	"	"	"	"	"	"	23-25	29-30			
16	"	22½-24	18	19	"	"	24-25	"	"			
17	35-37½	"	18-18½	"	"	"	"	24-25	30			
18	37½-40	"	18-19	"	17½-18	"	"	25-26	29-30			
19	"	22½-23½	18½-19	18½-19	"	"	"	"	"			
20	"	22½-23	19-19½	19	"	"	"	"	30			
21	"	22-22½	"	"	18-18½	22½-24	"	"	"			
22	33-35	21-22	19½	"	"	"	"	"	"			
23	32½-33	20-21	20	"	"	"	22-23	"	"			
24	32½	19-20	20-21	"	"	"	"	"	30-31			
25	"	"	"	"	18	23-24	"	"	"			
26	32 1-2-33	"	21	18½-19	"	"	"	"	31-32½			
27	"	18-19	22-23	"	"	24	"	"	"			
28	"	"	22½-23	18-19	17½-18	"	22-22½	"	"			
29	33-35		22-22½	"	"	24-25	"	26-27	"			
30	"		"	18-18½	"	"	"	"	"			
31	"		"	"	"	"	"	"	"			

THE POSSIBILITIES.

That there is room for great development in the poultry industry in this State will be apparent from a study of the conditions of the industry and of the markets in Oregon and the Pacific states. There is a large and stable market right at home, with the demand increasing faster than the supply. This is shown by the increase in prices during the past few years and by the increase in importations of eggs and poultry from other states. There is but little doubt that the farmers of the state could double their flocks of poultry and double the output of poultry products without seriously, if at all, affecting the prices. The profits of course do not depend upon the markets altogether. The markets may be good but the egg yield must also be good otherwise the flock will be kept at a loss.

I wish to speak here of the possibilities. It is possible to get an egg yield of 150 eggs per fowl per year, or even better than that. But that is more than twice as many eggs as the census gave the Oregon hens credit for laying. If those eggs can be sold at an average price of 20 cents a dozen they would bring $2.50. It is possible to do better than that. If a large proportion of the eggs were laid in winter the average price would be higher, but allowing for a fair average winter yield the price of 20 cents may not be too low.

As to the cost of production, eggs were produced at the Utah Experiment Station, when food was cheaper than it is now however, for less than five cents a dozen. The fowls consumed 60 cents worth of food during the year and laid more than an average of 150 eggs each. This was under ideal conditions probably, and it could hardly be expected with present market prices for food. It is difficult to get actual figures from poultrymen. Very few of them keep any records of egg yield or of food cost, but Mr. Joseph Schulte of Marion County, has given me a record of egg yield and of prices received and I quote his statement to show what has actually been done in this State in practical poultry keeping. The statement includes the monthly egg record, and the price received each month for the eggs for one year. The number of hens he states was between 275 and 300 which makes an average of over 150 eggs each.

1906	Eggs Laid	Price per Dozen	Amount Received
August	2,824	24 cts.	$ 56.48
September	2,532	26 "	54.86
October	2,135	30 "	53.38
November	1,093	36 "	32.79
December	2,215	35 "	64.40
1907			
January	2,817	34 "	95.75
February	4,735	25 "	97.81
March	6,221	20 '	103.70
April	6,073	16 "	80.96
May	6,015	17 "	85.21
June	4,320	22 "	79.12
July	4,144	22 "	75.97
Total	45,124		$880.66

It is seen that the most profitable month was March when eggs were 20 cents a dozen and the least profitable was November when they reached their highest figure or 36 cents. In other words as the prices advanced the returns decreased. The three most profitable months were January, February and March. It is possible by breeding f.om early layers and hatching early in the season to get an earlier egg yield in the fall months. In this way the receipts would be larger. The fact must be understood, however, that spring is the natural laying and breeding season of the fowls and that it will probably always be more difficult to get a good egg yield in the fall than in the spring. Fall and winter egg production is an artificial condition, and while the hens will lay with very little care in the spring it requires skillful handling to get eggs at any other season of the year. But it is this extra care or skill that will bring the poultryman his largest profits, Mr. Schulte could give no accurate statement of the amount of food consumed, but there was certainly a good margin of profit over and above cost of food. At the Maine Experiment Station where prices of foods are higher than here the cost of feeding Plymouth Rocks was $1.45 a year. At the Utah Station Plymouth Rocks were fed at less than $1.00 per fowl when foods were cheaper than now. Mr. Schulte's fowls were not confined to yards and were able to secure food on the range which would considerably lessen the cost, and $1.00 per fowl would be a very liberal allowance for food. That would leave $1.93 profit on the food consumed, or a profit on food of $579 for 300 hens.

Food Cost.—On farms that grow grain and where there is much waste grain that the chickens eat the cost would be considerably less. It is doubtful if on the average farm the actual outlay for food will exceed fifty cents a year per fowl at present prices for grain. Where the food has to be all purchased and paid for at market prices the cost will vary from $1.00 to $1.25 per fowl. Outside of food and labor the expenses are not great. The cost of raising a pullet will usually be offset by the price received for the hen when she is marketed, and with good management a profit will be made on the sale of the surplus cockerels.

Labor Cost.—The cost of labor is a more uncertain item. On the general farm where 50 or 100 hens are kept the labor item may be practically nothing, but where poultry-keeping is made a regular occupation the labor must be taken into account. There must be good management or the cost of labor will eat up the profits. It is not a question altogether of getting the hens to lay; it is a question of economy in production. Enough labor is frequently expended on 100 hens to keep 1000 hens. To reduce the labor to an economical basis the feeding and general care of the fowls must be systematized so that the poultryman may take care of a large number of fowls. A minimum number of fowls for one man would be about 1000 hens. A modest living may be made on half that number, but one man should be able to take care of at least 1000 hens, and those hens should bring in a revenue of $1,000 to $1,500 a year above cost of food. It is possible by keeping a larger number and employing extra help to do better than that. On a large poultry farm with arrangements in housing and methods of feeding that will permit of the proper economy in labor a flock of 5000 hens may be taken care of by a man and a boy. On such a farm a boy of 15 or 16 years of age will be of about as much service as a man. This is actually being done. On a farm in Connecticut which I visited some 18 months ago, owned by G. G. Tillinghast of Vernon, I found the labor problem under such control that Mr. Tillinghast said one man could take care of 5000 laying hens. Where the hatching and brooding is done on the farm some extra help will be necessary. On no place that I have been is this accomplished where the colony system of housing is not used, except possibly in one or two cases, one being the Lakewood Poultry Farm of New Jersey, where continuous stationary houses are in use. On this farm Mr. Brown, the manager, said to me: "I, with the help of a boy, do all the work connected with the hatching, caring for and feeding the young stock and tending to the brooders. Another man and boy do all the work in caring for the laying hens." The stock numbered about 5000 hens. Another successful poultry farm which I visited where long stationary houses are used is near Bangor, Maine. On this farm, which is owned by Professor Gowell of the Maine Experiment Station, is one continuous house 400 feet long and 20 feet wide, in which 2000 Plymouth Rocks are kept. It is divided into pens 20x20 feet and 100 fowls are kept in each. Mr. Gowell informed me that one man did all the work of caring for this flock except every other Saturday when the straw bedding was removed. On the large poultry farms of California one man usually does the work required in taking care of four to five thousand laying hens.

Profits From Large Flocks.—As to the profits from large flocks there is a diversity of opinion. One dollar per hen above cost of food with good stock and good management ought to be a minimum of profit per year. Experienced poultrymen in California told me that it could not be done, others that better than that could be done. The truth is that few if any of them know exactly what their flocks are returning. The item of loss in the stock is undoubtedly large on most farms, and this is not surprising from the crude methods that prevail generally. The fact that poultry-keeping there is engaged in to the exclusion of almost every other industry is pretty good proof that there is money in it. When a poultryman with two or three hundred acres of land on which nothing is grown except chickens and every pound of food used by the hens is purchased and hauled some ten miles, and that at the slack season of the year the owner and his wife can take a pleasure trip across the continent, visiting the Yellowstone National Park on the way, as I found was being done, is another pretty clear indication that some men make poultry-keeping pay. If the chickens were not a paying proposition you would hardly expect a poultryman to indulge in the luxury of an automobile, and I found a poultryman near Petaluma with an automobile, and there may have been others. The fact that the poultrymen in the neighborhood of Petaluma sold $1,500,-000 worth of poultry and eggs in 1906 is pretty good evidence that there is money in it. The town is supported very largely by the poultry business. The largest merchant there is a dealer in poultry foods and supplies. I was informed that the Lakewood Poultry Farm already mentioned made a clear profit of $7,000 in 1905. Professor Gowell is undoubtedly clearing nearer $2 per hen than $1. Mr. Tillinghast of Connecticut was rated worth $60,000, made very largely from his chickens.

These are a few instances of success. No doubt there are more failures than successes, but the few successes prove that the business is a profitable one. It is folly to start in the business on a large scale without previous experience. Mr. E. B. Thompson, a noted poultry breeder in New York State, gave me the following as his opinion: "A man without experience who goes into the poultry business and puts all his money into it, it may be several thousand dollars, will lose all he has." But that is not the fault of the business; it is the misfortune of the man. It is not necessary to keep a thousand hens or 5000 hens to make a living out of them. There are numbers who are making a modest living on less than 1000 hens. To cite a local case, Mr. William Rogers of Grants Pass is making a living for himself and wife with about 500 hens on twelve acres of land. Mr. Ringhouse of Gresham believes that a thousand dollars profit can be made above cost of food with 1000 hens and with extra care, good stock and good management as much as $2,000 may be made.

Poultry and Mixed Husbandry.—What has been said of the possibilities and the profits, applies only to special poultry farming or poultry-keeping as a special business. Larger profits are undoubtedly possible on the farms of mixed husbandry where the feed item and the labor item, as well as the item of land, are of smaller account than on the special poultry farms. There are great possibilities of development in poultry-keeping on the farm, where the farmer gives intelligent care to the business. I think it is pretty well understood that the farmer gives very little care generally to this branch of mixed husbandry. Judging from the way they are handled it looks as though chickens on the farm are tolerated as a sort of necessary evil. It is not true of all farmers of course, otherwise the farms of the country would not be producing the tremendous annual crop of poultry and eggs that they are now doing. But the majority of farmers give less care to the poultry than they do to any other part of the farming operations. If the farmer would sometimes stop to think and do a little figuring he might possibly discover that the chickens were the best paying part of the farm.

Larger profits are undoubtedly made on the farm of mixed husbandry than on the poultry farms, because the labor item and the feed item are smaller. It has been demonstrated in Rhode Island that flocks of a thousand hens or more may be profitably kept on general farms. In a section of that State for about 60 years poultry-keeping has been one of the principal branches of general farming. On many farms it is now the main issue. From a thousand to two thousand hens are frequently kept and the plan is to keep them in the colony houses, change them on to a fresh portion of the farm each year and hatch the chickens by natural methods. It is the usual plan for chickens and cattle or sheep to range over the same pasture fields, and in that way two crops are obtained from the same ground.

A system of this kind opens up vast possibilities to many of our own farmers. By fencing off portions of the farm with wire netting fences the poultry would not interfere with the raising of grain or other crops on other portions of the farm. They would be a positive help to the land in many ways. Their droppings would fertilize the land—and hen manure is richer than that of any of the domestic animals—and they would be of great service in riding the land of grasshoppers and other injurious insects. When the grain crops are harvested the poultry could be given the liberty of the stubble fields where during the fall months they would find practically all the food required. Poultry-keeping in this way could be made an important feature of a practical system of rotative farming.

CLIMATE AND POULTRY-KEEPING.

Climate is not so much a factor in poultry-keeping as in most other branches of agriculture. Poultry is found in every state of the Union and probably in every county. That the climate of Oregon is not unfavorable for poultry-keeping it is only necessary to study the statistics of production given on a preceding page. It is of course worth considering by the man looking for a location whether western Oregon with its open winters and freedom from snow and zero temperatures does not offer opportunities for the production of eggs and poultry that are not found in eastern and middle west states. That poultry thrive in cold sections where snow and zero weather prevail is not to be denied, but the labor and expense of caring for them is undoubtedly greater there. To secure an egg yield in winter where the climate is severe entails more expense for housing and more care in the feeding. It is probably true that the smallest profits are made during the winter months though the prices are very much higher than in spring and summer, because the egg yield is so small from the average flock as to leave little or no margin of profit. It is also true that the egg yield is quickly affected by changes in the weather especially in the temperature. A sudden change from mild to cold weather means a certain check to egg production and though the weather soon moderates it will often take several weeks before the egg yield gets back to where it was. The only way to prevent this is to provide housing that will protect the fowls from too sudden changes in temperature. This entails more expense in housing and consequently diminished profits, but what is of more importance is the highly artificial conditions that it necessitates. Such artificial conditions in housing and temperature may for a time increase the egg yield slightly, but it has never yet been shown that domestic poultry so-called can be kept under highly artificial conditions without a lowering of the vitality of the flock. It is a question with poultry-men in eastern states whether it pays to attempt to force the egg yield in winter by confining the fowls in close warm quarters.

It would appear therefore that there are certain advantages that this State possesses over sections of the country where zero weather and snow prevail. First, a milder climate and less severe changes in temperature than is characteristic of eastern states. Second, in sections of the State with no snowfall the poultry can range over the fields and find animal food and green food which are often hard to get where the snow covers the ground.

The heavy rainfall of western Oregon and small percentage of sunshine may be set down as a disadvantage, but when the nature of the rainfall is understood it is doubtful whether it is very much of a detriment. Owing to the moderating influence of the Pacific Ocean these rains are warm and have not the chilling effect of the rains in eastern states. The temperature of western Oregon in the winter months is usually higher when it rains than when the sky is cloudless, and the fowls will usually be found out in the rain except when it is very heavy which is not often the case. One poultryman in Marion County said to the writer in November, before the rainy season set in, that he wished it would rain because he said his hens laid better when it rained. The explanation of this, if it is true, may not be in the rain itself but in the fact that it brings to the surface many angleworms which supply the lack of animal food in the ration. During the rainy season of winter large numbers of these worms are to be found.

Turkeys are successfully raised in Oregon, and turkeys are known to be easily affected by rain, but the fact that the rains are warm no doubt largely accounts for the success in turkey raising in this State. Douglass county in Oregon produces several times more turkeys than the state of Rhode Island, noted for turkeys.

Another thing in favor of the mild climate and freedom from snow is that the fowls are able to secure practically the year round all the green food necessary in the fields. And finally, the fowls in their search for food in the fields get the exercise which is necessary for the maintenance of health.

It is worthy of mention in this connection that the largest special poultry district in the United States is found in northern California that has no snowfall. That district is somewhat similar to that of western Oregon with its open winters, mild and humid climate and nearness to the ocean. But the rainfall is slightly heavier in Oregon and the temperature somewhat lower.

My investigations of the poultry industry of Oregon have been confined to the western part of the State, the region west of the Cascade mountains. This section at the present time produces more poultry products than the larger area of the state east of the Cascades. As the agriculture of central and eastern Oregon becomes developed we may expect greater development of poultry-keeping, and probably in time that great agricultural area may produce more poultry products than the older section of the State in western Oregon. The climatic conditions are different east of the mountains, the heavy rainfall is absent and snow covers the ground during part of the winter. The climate there is more characteristic of the Rocky Mountain region, though no such severe weather prevails as in the middle west and northwestern states. If it should prove that a dry climate with plenty of sunshine but lacking the severe winter changes of temperature of the east is the ideal one for poultry, we may expect a great growth of the poultry industry east of the Cascades in Oregon. Undoubtedly on the grain ranches of central and eastern Oregon where food is cheap there is opportunity for great profit in poultry raising.

SOILS AND POULTRY-KEEPING.

The question is often asked what is the best location or site for poultry-keeping. The soil has undoubtedly to be taken into consideration. It has a bearing on the question of maintaining the health of the fowls. The soil should be well drained and porous, and the full importance of this is not always realized. It is more important in a humid region than in a dry one. Where there is much sunshine there is less contamination of the soil. It is well known that sunshine is a germ destroyer, and disease germs find in a wet poorly drained soil better conditions for development than in a dry one.

This feature of our climate must be recognized. It is the one thing above all others that we have got to take into account in this country of mild, open, wet winters. While I believe we have on the whole a decided advantage over the eastern states for profitable poultry keeing, we must nevertheless face the fact that disease-producing germs in the poultry yards have here a longer season of activity than in regions where the snow covers the ground for a portion of the year. This is the one thing of any consequence as I look at it where the conditions of western Oregon are not as good as in most of the eastern states. While snow on the ground is a disadvantage in poultry keeping, it would be a decided advantage if there were no other way of keeping the fowls from a soil contaminated with disease-producing germs. This danger of soil contamination or ground poisoning in the humid regions of Oregon has got to be met or there are liable to be serious outbreaks of poultry diseases. Where intensive poultry culture prevails in the East there are not lacking cases where the poultryman has been driven out of business because of ground poisoning. This intensive poultry-keeping has resulted in a general lack of vigor in the stock. The effects are shown in reduced egg yield, in lack of fertility and weakness of embryo in the egg, in small hatches and in high mortality in the chicks. A case has come to notice of a poultryman near Boston who in former years conducted a profitable poultry business but in late years he found it practically impossible to raise the chicks after being hatched, and he ascribed the trouble to contamination of the ground. If that would occur in the eastern states where the disease producing germs are inactive a large part of the year the probability is that in the western part of Oregon with its open winters and humid climate greater care will be necessary to prevent catastrophies of this nature. In the Willamette Valley we have a climate more like that of Great Britain than of the eastern states, and I wish to quote the statement made by Professor Edward Brown of Reading College, England, whom I had the pleasure of meeting in New York State, when he visited this country to investigate and report upon the poultry conditions in America. He said: "Were the conditions in America as to climate the same as in Britain, I should have anticipated ere this a great outbreak of epidemic diseases of one form or another." This is the statement of a man who has made poultry-keeping a life study and has studied it in nearly all the civilized countries on the globe. The reason he made the statement was, as he explained, that the importance of fresh ground or cultivation had not been recognized in America.

We must therefore recognize the fact that in western Oregon the climatic conditions are not favorable to the destruction of disease-producing germs. Fortunately this is a condition against which poultrymen are not helpless. Rational methods will overcome it.

Land that is well drained and porous should be selected if possible. An open porous soil will be less subject to ground poisoning than a soil that is clayey in its nature and underlaid with hardpan. A soil with a subsoil of hardpan should be avoided if at all possible and the poultryman should never buy a piece of land for a poultry farm until he has dug down into the soil and has learned its nature. Rains as they fall on a light porous soil will quickly disappear and carry with them much of the filth on the surface. This will postpone but will not prevent the inevitable poisoning of the ground no matter what may be the conditions of soil or climate. A light porous soil will not so readily become contaminated as a heavy clay soil, but it will in time.

Professor Pernot, in Bulletin 64, discussing tuberculosis in fowls, states that he found in the evacuations of a single specimen countless numbers of the tubercle bacilli. This shows how easily and quickly the ground may become infected from one diseased bird, and how important it is that the ground be kept clean

What are the remedies? What are the conditions that must be observed if the ground poisoning is to be prevented? The first and the easiest way to overcome it is to have enough land so that the chickens may be given fresh ground to range over every year; and better, change the houses on to fresh ground every month. Where the land is limited and it is necessary to confine the fowls in yards there should be two yards for each pen of fowls so that the yards may be cultivated and cropped every other year. Keeping the fowls on a small piece of clean ground is better than letting them run on double the area of filthy impure ground. The larger the runs of course the better for the fowls, but it is a question of choosing the lesser of two evils. With careful feeding and management they may be profitably kept on a small piece of ground if kept free from disease-producing filth, but keeping them on the same ground year after year without any systematic cultivation or purifying of the soil and expecting to avoid trouble from diseases will result in certain disappointment. That poultry-keeping in Great Britain is not subject to epidemics of diseases is largely due to the fact that the colony system of housing the fowls prevails there. The same thing is true of Petaluma, California, where the climate is also mild and humid. The most successful poultrymen there are several miles from town where the fowls have unlimited range and are kept in movable colony houses.

On the general farm where the fowls have the liberty of the fields there will be little danger from impure ground especially if the houses are moved frequently. As a general rule poultry may be successfully kept on most kinds of soil but the heavy clay soils will require more careful handling than the lighter soils. On the heavy soils there is greater need that movable houses be used but where this is not possible the ground should be frequently cultivated and cropped, and in places underdrainage may be necessary.

The climate and soil of Oregon is as good as any in the world for poultry-keeping, and it is doubtful if there is any other agricultural industry in the State of Oregon that offers the same opportunities for profit-making as poultry-keeping, yet we have local conditions not altogether favorable that we must meet. Eastern conditions do not apply, Eastern methods will not always do, and we must remember that our open mild wet winters which though favorable to the production of eggs bring with them dangers which are not experienced in the East, and only intelligent methods will overcome them.

METHODS OF HOUSING.

During the past few years the tendency among poultrymen has been to build houses on a plan to admit of an abundant supply of fresh air. More consideration is now being given to fresh air than warmth in poultry houses. The mistake in the past has been in a failure to recognize the fact that the housing of poultry is an artificial condition, for our domestic poultry still retain considerable of the wild nature. This is shown when fowls persist in flying into the trees to roost in preference to seeking the shelter of houses provided for them.

It has never yet been demonstrated that fowls can be kept in warm houses without injury to the constitutional vigor of the flock, and the vigor of the flock must be the first concern of the poultry breeder.

The difficulty is in building a house that will not be subject to too great variations in temperature between night and day and at the same time well ventilated, and not too damp, without furnishing artificial heat. It is possible by artificially heating the house to get a slightly better egg yield, as is indicated by experiments at some of the Stations, but they also indicate that the increased yield was at the expense of vitality of the fowls. It is very important that the temperature does not vary too much in the house. In a warmly built house with windows in the south the temperature will be high during the day when the sun shines and at night it will be very low, the glass permitting the escape of the heat as rapidly as it enters. This is because of the chilling effect on the fowls when subjected to extreme variations of temperature; and second, because of the dampness it produces in the house. During the day the warm air in the house is taking up the moisture and at night, owing to the great fall in temperature, the air becomes very humid, often totally saturated as is seen when moisture condenses on the walls. It is not dampness on the walls, however, that is bad, it is the moisture in the air; but the dampness on the walls is a sure indication that the air is as damp or humid as it is possible to to be. Chickens would be better roosting on the tree than in a house where such conditions prevail. By taking out the windows altogether these conditions would be improved. They would be still futher improved by cutting out a larger portion of the front, and this would make what is generally known as an open-front or fresh-air house. The open front may be adapted to a small house or to a large one, a colony house or a stationary house. The open-front house has been tried in all sections of the country, in the cold as well as in the warm. In cold sections some poultrymen cover the opening with a curtain of ordinary muslin, and then it is called a curtain-front house. This shuts out the storms and retains some of the heat in the house. As the material is thin there is a free circulation of air but without drafts, and the temperature of the house is nearly that of out doors. The curtain-front house has been thoroughly tested at the Experiment Station of Maine and appears to meet the conditions there better than any other house and gives good satisfaction in the long stationary house. It has been tried with the colony house at the Utah Station with good results, though the temperature went to 13 degrees below zero and snow was on the ground about two months during the winter. It was found that the fowls maintained a greater weight in the colony open-front house than in the warm closed house.

It is a question with many whether the curtain is necessary. In the milder portions of this State there is certainly less need for it than in the eastern states, if there is any advantage at all in it. In western Oregon there is no snow to drift into the house and there is not so much wind as in eastern states. This question needs further investigation but in the meantime the writer doesn't hesitate to recommend the open-front house. The curtain however entails small expense and there may be seasons and places where it would be an advantage to use it.

Movable Or Stationary Houses.—Having decided on the fresh air house, whether it be an open-front or curtain-front house, the next question is shall it be a movable colony house or a stationary house. This has been referred to on a previous page in discussing soils and ground poisoning. Where enough land is available the movable colony house is without question the safest and most practicable house to use. On small pieces of ground or in suburban poultry-keeping, the stationary house must necessarily be used.

I find that in this State most of the leading poultrymen are building the colony house with the open front. Mr. H. Ringhouse of Gresham has adopted a combination of the colony and stationary house, but in each case the open-front small house is used. His plan is to keep his breeding stock, from which he secures eggs for hatching, in movable colony houses while his general laying stock will be kept in the stationary houses. (Figs. 1 and 2). The dimensions of the stationary houses are 10 x 12 feet, and fifty and sixty birds were successfully kept in some of them, but he aims to keep about 30 in each. The opening in front is 4 x 8 feet and the house is floored. He said there were no drafts in the house and the wind would not blow out a candle on a windy night where the fowls roosted. Mr. Ringhouse is building his movable houses smaller, the dimensions being 6 x 10 feet.

The same type of open-front house is being built by Mr. Reynolds on the Krebs hop farm at Brooks, and by Mr. E. K. Brown at Eugene. Mr. Brown also uses these houses in the hop yards. The plan is to move them between the rows of hops. Mr. Reynold's houses are 6 x 10 feet while Mr. Brown's are 6 x 12 feet. Both houses have floors. (Figs. 6 and 12). A very serviceable and cheap house is shown in Fig. 7. Mr. Brown of Eugene has built a number of these. They are made with two piano boxes, back to back, with the boards of the back removed. The houses are covered with a good quality of roofing paper. These different houses are all closely built on three sides so as to avoid any possibility of drafts on the fowls.

The Colony House.—In Sonoma County, California, where specialized poultry-keeeping is engaged in more extensively than in any other section of the United States, the prevailing plan is the movable colony house. The town of Petaluma shipped in 1906 some one million and a half dollars worth of poultry and eggs and the system that has produced these results is worthy of careful study. The style of house is almost invariably the colony house. Near the town on the smaller "ranches" the houses are not movable, but out a few miles on the larger farms they regularly moved, I was informed, once a month. Fig. 4 gives a view of one of the largest poultry farms in that section, belonging to Mr. W. Freeman, where I was told some 8000 laying hens are regularly kept. The photograph shows how the houses are scattered widely apart over the farm.

Figs. 19 to 22 show the general character of the houses. The prevailing house is about 7 x 14 feet some a little larger. They are very cheaply built and little pretense is made to shut out the drafts, very few of the houses being battened. The ventilation, where there is any provided except the cracks in the walls, is admitted above the door, as will be seen by referring to Fig. 20. The houses are without floors.

The plan of colonizing generally prevailing on the large poultry farms of California is to keep three houses together, the center one being the laying house and the other two the roosting houses Usually 200 fowls are kept in a colouy, 100 for each roosting house. Such crowding of poultry flatly contradicts all preconceived views of poultrymen thoughout the country. "Do not crowd poultry," has been a theme on which poultry writers have written with strong convictions, but we may well hesitate to criticise a plan which has been productive of greater practical results probably than any other system in the country. At the same time I would hesitate to advise building a house through which the wind whistled from all sides and in which every square foot of floor space was utilized for roosting room. On many farms the allottment of floor space did not exceed one square foot per fowl.

The question of floor space, however, is of little importance; it is more a question of fresh air. With plenty of pure air fowls will stand considerable crowding. About every farmer I interviewed there stated that the losses they had from deaths in the flocks were mostly due to so-called roup. This is a pretty strong indication that the system is not ideal. These losses may be due to several causes: first, lack of inherent vitality of the fowls; second, drafts in the house; third, crowding, producing an overheated condition and subsequent chilling when let out on a cold morning. Only on one farm did I find an open-front house and it was giving good satisfaction to the owner. Most of the farmers with whom I talked had not given the subject of open front houses any consideration, evidently preferring to follow in the beaten path.

In Rhode Island there is another successful poultry district, probably second only to Petaluma in the extent of its poultry business; and curiously enough the houses there are also of a crude pattern. In a recent letter to me Professor Graham of the Storrs Agricultural College, Connecticut, speaks of these houses as follows: "The houses are made of unmatched board, and many do not even have the cracks covered. Their whole methods are very crude. They have any quantity of roup but claim hens die of practically nothing else in that locality."

While great credit is due to the poultrymen of Rhode Island and California, it must not be concluded that cracks are necessary in poultry houses, but the system is at least a strong protest against the practice of putting hens in air-tight houses with the idea of making them comfortable.

Large vs. Small Stationary Houses.—Where land is more restricted and where the fowls have to be confined in yards, the stationary house must be adopted. Shall it be one long house or several small ones? The argument that the long, continuous house is cheaper to build than several small ones of similar capacity does not hold true in reality since it has been shown that fowls cannot be kept in one long house without partitions to stop the drafts. The cost of the partitions will about equal the cost of making an extra number of ends to small houses. But the chief argument in favor of the small house is that it is possible to provide a system of cultivation where the houses are small and set far enough apart to permit of double yards. It has been pointed out on another page in discussing tainted soil, that it is very important that double yards be provided where stationary houses are used so that the soil may be cultivated every alternate year. The plan of stationary houses as adopted by Mr. Ringhouse, shown in Figs. 1 and 2, is to be highly commended. These houses are provided with double yards which are wide enough to admit of thorough cultivation.

There are many examples throughout the country where the long, continuous house has failed to give satisfaction. Mr. F. L. Miller, of Corvallis, after one year's trial of a continuous house 800 feet long, was so fully convinced that the principle was wrong that he has taken the house down and built small stationary ones (See Figs. 12 and 13). The houses which he has built are 16 x 24 with an open-front 3 x 18.

Mr. J. Lenz, of Hillsboro, is building quite an extensive plant on 15 acres of land. He is now building eight houses which are 14 x 40 feet (Fig. 5). He aims to keep sixty fowls in each house, which is a very liberal allowance of space. The house is floored and one end of it is utilized for a roosting place and the other end as a scratching floor where the fowls will find shelter during stormy days. The house is 7 feet high at the front and 5 feet high at the rear. There are six windows, 3 feet square, in each house. The houses are set two feet above the ground. This is to afford shelter from storms as well as to furnish shade for the chickens, but the principal idea in view was the control of the rat pest. Last year Mr. Lenz lost 500 chicks from weasels and rats and in building the new houses he built them high so there would be no space under the floor in which the vermin could safely harbor. One view of Fig. 5 shows the incubator house and brooder house Mr. Lenz has built; the former is 16 x 30 feet and the latter house 14 x 120 feet.

Mr. Joseph Schulte, of Marion County, is getting practical results from housing 300 laying hens in one continuous house, about 90 feet long, divided into alternate roosting and scratching sections. The fowls are not yarded but run together and have the liberty of the fields. The returns he is getting from his flock are given on another page.